CODEX ALIMENTARIU

Waters

First edition

WORLD HEALTH ORGANIZATION

FOOD AND AGRICULTURE ORGANIZATION OF THE UNITED NATIONS

Rome, 2007

ISBN 978-92-5-105836-7

THE CODEX ALIMENTARIUS COMMISSION

The Codex Alimentarius Commission is an intergovernmental body with over 170 members, within the framework of the Joint FAO/WHO Food Standards Programme established by the Food and Agriculture Organization of the United Nations (FAO) and the World Health Organization (WHO), with the purpose of protecting the health of consumers and ensuring fair practices in the food trade. The Commission also promotes coordination of all food standards work undertaken by international governmental and non governmental organizations.

The *Codex Alimentarius* (Latin, meaning Food Code) is the result of the Commission's work: a collection of internationally adopted food standards, guidelines, codes of practice and other recommendations. The texts in this publication are part of the Codex Alimentarius.

WATERS
First edition

Codex standards for waters and codes of hygienic practice are published in this compact format to allow their wide use and understanding by governments, regulatory authorities, food industries and retailers, and consumers. This first edition includes texts adopted by the Codex Alimentarius Commission up to 2007.

Further information on these texts, or any other aspect of the Codex Alimentarius Commission, may be obtained from:

The Secretary
Codex Alimentarius Commission
Joint FAO/WHO Food Standards Programme
Viale delle Terme di Caracalla
00153 Rome, Italy
Fax: +39 06 57054593
E-mail: codex@fao.org
http:// www.codexalimentarius.net

WATERS
First edition

CODEX STANDARD FOR NATURAL MINERAL WATERS

CODEX STAN 108-1981

1. SCOPE

This standard applies to all packaged natural mineral waters offered for sale as food. It does not apply to natural mineral waters sold or used for other purposes.

2. DESCRIPTION

2.1 Definition of natural mineral water

Natural mineral water is a water clearly distinguishable from ordinary drinking water because:

a) it is characterized by its content of certain mineral salts and their relative proportions and the presence of trace elements or of other constituents;
b) it is obtained directly from natural or drilled sources from underground water bearing strata for which all possible precautions should be taken within the protected perimeters to avoid any pollution of, or external influence on, the chemical and physical qualities of natural mineral water;
c) of the constancy of its composition and the stability of its discharge and its temperature, due account being taken of the cycles of minor natural fluctuations;
d) it is collected under conditions which guarantee the original microbiological purity and chemical composition of essential components;
e) it is packaged close to the point of emergence of the source with particular hygienic precautions;
f) it is not subjected to any treatment other than those permitted by this standard.

2.2 Supplementary definitions

2.2.1 Naturally carbonated natural mineral water

A ***naturally carbonated natural mineral water*** is a natural mineral water which, after possible treatment in accordance with Section 3.1.1 and re-incorporation of gas from the same source and after packaging taking into consideration usual technical tolerance, has the same content of carbon dioxide spontaneously and visibly given off under normal conditions of temperature and pressure.

2.2.2 Non-carbonated natural mineral water

A ***non-carbonated natural mineral water*** is a natural mineral water which, by nature and after possible treatment in accordance with Section 3.1.1 and after packaging taking into consideration usual technical tolerance, does not contain free carbon dioxide in excess of the amount necessary to keep the hydrogen carbonate salts present in the water dissolved.

Adopted 1981. Amendment 2001. Revisions 1997, 2008.

2.2.3 **Decarbonated natural mineral water**

A ***decarbonated natural mineral*** is a natural mineral water which, after possible treatment in accordance with Section 3.1.1 and after packaging, has less carbon dioxide content than that at emergence and does not visibly and spontaneously give off carbon dioxide under normal conditions of temperature and pressure.

2.2.4 **Natural mineral water fortified with carbon dioxide from the source**

A ***natural mineral water fortified with carbon dioxide from the source*** is a natural mineral water which, after possible treatment in accordance with Section 3.1.1 and after packaging, has more carbon dioxide content than that at emergence.

2.2.5 **Carbonated natural mineral water**

A ***carbonated natural mineral water*** is a natural mineral water which, after possible treatment in accordance with Section 3.1.1 and after packaging, has been made effervescent by the addition of carbon dioxide from another origin.

2.3 Authorization

Natural mineral water should be recognized as such by the responsible authority of the state, in which the natural mineral water has emerged.

3. COMPOSITION AND QUALITY FACTORS

3.1 Treatment and handling

3.1.1 Treatments permitted include separation from unstable constituents, such as compounds containing iron, manganese, sulphur or arsenic, by decantation and/or filtration, if necessary, accelerated by previous aeration.

3.1.2 The treatments provided for in Sections 2.2.1, 2.2.2, 2.2.3, 2.2.4, 2.2.5 and 3.1.1 above may only be carried out on condition that the mineral content of the water is not modified in its essential constituents, which give the water its properties.

3.1.3 The transport of natural mineral waters in bulk containers for packaging or for any other process before packaging is prohibited.

3.2 Health-related limits for certain substances

Natural mineral water in its packaged state shall contain not more than the following amounts of the substances indicated hereunder:

3.2.1	Antimony	0.005 mg/l
3.2.2	Arsenic	0.01 mg/l, calculated as total As
3.2.3	Barium	0.7 mg/l[1]
3.2.4	Borate	5 mg/l, calculated as B

[1] Pending further review of new scientific evidence by an appropriate scientific body to be determined by FAO/WHO.

3.2.5	Cadmium	0.003 mg/l
3.2.6	Chromium	0.05 mg/l, calculated as total Cr
3.2.7	Copper	1 mg/l
3.2.8	Cyanide	0.07 mg/l
3.2.9	Fluoride	See section 6.3.2
3.2.10	Lead	0.01 mg/l
3.2.11	Manganese	0.4 mg/l
3.2.12	Mercury	0.001 mg/l
3.2.13	Nickel	0.02 mg/l
3.2.14	Nitrate	50 mg/l, calculated as nitrate
3.2.15	Nitrite	0.1 mg/l as nitrite
3.2.16	Selenium	0.01 mg/l

The following substances shall be below the limit of quantification[2] when tested, in accordance with the methods prescribed in Section 7:

3.2.17 Surface active agents[3]

3.2.18 Pesticides and PCBs[3]

3.2.19 Mineral oil[3]

3.2.20 Polynuclear aromatic hydrocarbons[3]

4. HYGIENE

4.1 It is recommended that the products covered by the provisions of this standard be prepared in accordance with the applicable sections of the *Recommended International Code of Practice – General Principles of Food Hygiene* (CAC/RCP 1-1969), and in accordance with the *Recommended International Code of Practice for the Collecting, Processing and Marketing of Natural Mineral Waters* (CAC/RCP 33-1985).

[2] As stated in the relevant ISO methods.

[3] Temporarily endorsed pending elaboration of appropriate method(s) of analysis.

4.2 The source or the point of emergence shall be protected against risks of pollution.

4.3 The installations intended for the production of natural mineral waters shall be such as to exclude any possibility of contamination. For this purpose, and in particular:

a) the installations for collection, the pipes and the reservoirs shall be made from materials suited to the water and in such a way as to prevent the introduction of foreign substances into the water;
b) the equipment and its use for production, especially installations for washing and packaging, shall meet hygienic requirements;
c) if, during production it is found that the water is polluted, the producer shall stop all operations until the cause of pollution is eliminated;
d) the observance of the above provisions shall be subject to periodic checks in accordance with the requirements of the country of origin.

4.4 Microbiological requirements

During marketing, natural mineral water:

a) shall be of such a quality that it will not present a risk to the health of the consumer (absence of pathogenic microorganisms);
b) furthermore it shall be in conformity with the following microbiological quality specifications:

First examination		Decision
E. coli or thermotolerant coliforms	1 × 250 ml	must not be detectable in any sample
Total coliform bacteria	1 × 250 ml	
		if ≥ 1 or ≤ 2 ⇒ a second examination is carried out
Fecal *streptococci*	1 × 250 ml	
Pseudomonas aeruginosa	1 × 250 ml	
		if > 2 ⇒ rejected
Sulphite-reducing anaerobes	1 × 50 ml	

Second examination	*n*	*c*[4]	*m*	*M*
Total coliform bacteria	4	1	0	2
Fecal *streptococci*	4	1	0	2
Sulphite-reducing anaerobes	4	1	0	2
Pseudomonas aeruginosa	4	1	0	2

[4] Results of the first and second examinations.

Second examination shall be done using the same volumes as for the first examination.

n: number of sample units from a lot that must be examined to satisfy a given sampling plan.

c: the maximum acceptable number, or the maximum allowable number of sample units that may exceed the microbiological criterion m. When this number is exceeded, the lot is rejected.

m: the maximum number or level of relevant bacteria/g; values above this level are either marginally acceptable or unacceptable.

M: a quantity that is used to separate marginally acceptable quality from unacceptable quality foods. Values at or above M in any sample are unacceptable relative to either health hazard, sanitary indicators, or spoilage potential.

5. PACKAGING

Natural mineral water shall be packed in hermetically sealed retail containers suitable for preventing the possible adulteration or contamination of water.

6. LABELLING

In addition to the *Codex General Standard for the Labelling of Prepackaged Foods* (CODEX STAN 1-1985), the following provisions shall apply:

6.1 The name of the product

6.1.1 The name of the product shall be "*natural mineral water*".

6.1.2 The following designations shall be used in accordance with Section 2.2 and may be accompanied by suitable descriptive terms (e.g., still and sparkling):

- Naturally carbonated natural mineral water;
- Non-carbonated natural mineral water;
- Decarbonated natural mineral water;
- Natural mineral water fortified with carbon dioxide from the source;
- Carbonated natural mineral water.

6.2 Name and address

The location of the source and the name of the source shall be declared.

6.3 Additional labelling requirements

6.3.1 **Chemical composition**

The analytical composition giving characteristics to the product shall be declared in the labelling.

6.3.2 If the product contains more than 1 mg/l of fluoride, the following term shall appear on the label as part of, or in close proximity to, the name of the product or in an otherwise prominent position: "*contains fluoride*". In addition, the following sentence should be included on the label: "*The product is not suitable for infants and children under the age of seven years*" where the product contains more than 1.5 mg/l fluorides.

6.3.3 If a natural mineral water has been submitted to a treatment in accordance with sub-section 3.1.1, the result of the treatment shall be declared on the label.

6.4 Labelling prohibitions

6.4.1 No claims concerning medicinal (preventative, alleviative or curative) effects shall be made in respect of the properties of the product covered by the standard. Claims of other beneficial effects related to the health of the consumer shall not be made unless true and not misleading.

6.4.2 The name of the locality, hamlet or specified place may not form part of the trade name unless it refers to a natural mineral water collected at the place designated by that trade name.

6.4.3 The use of any statement or of any pictorial device which may create confusion in the mind of the public or in any way mislead the public about the nature, origin, composition and properties of natural mineral waters put on sale is prohibited.

7. METHODS OF ANALYSIS AND SAMPLING

See relevant Codex texts on methods of analysis and sampling.

GENERAL STANDARD FOR BOTTLED/ PACKAGED DRINKING WATERS (Other than Natural Mineral Waters)

CODEX STAN 227-2001

1. SCOPE

This Standard applies to waters for drinking purposes other than Natural Mineral Waters, as defined in the Revised Codex Standard CODEX STAN 108-1981, that are prepackaged/bottled[1] and are suitable for human consumption.

2. DESCRIPTION

2.1 Packaged waters

"Packaged waters", other than natural mineral waters, are waters for human consumption and may contain minerals, naturally occurring or intentionally added; may contain carbon dioxide, naturally occurring or intentionally added; but shall not contain sugars, sweeteners, flavourings or other foodstuffs.

2.1.1 Waters defined by origin

"Waters defined by origin", whether they come from the underground or from the surface, defined under the present standard share the following characteristics:

a) they originate from specific environmental resources without passing through a community water system;

b) precautions have been taken within the vulnerability perimeters to avoid any pollution of, or external influence on, the chemical, microbiological and physical qualities of water at origin;

c) collecting conditions which guarantee the original microbiological purity and essential elements of their chemical make-up at origin;

d) from the microbiological standpoint, are constantly fit for human consumption at their source and are kept in that state with particular hygienic precautions until and while packaging in accordance with provisions of sections 3 and 4;

e) are not subject to any modification or treatment other than those permitted under Section 3.1.1.

2.1.2 Prepared waters

"Prepared waters" are waters that do not comply with all the provisions set for waters defined by origin under subsection 2.1.1. They may originate from any type of water supply.

---

[1] As defined in *Codex General Standard for the Labelling of Prepackaged Foods:* "prepackaged foods to be offered as such to consumer or for catering purposes".

3. ESSENTIAL COMPOSITION AND QUALITY FACTORS

3.1 Modifications and handling of packaged waters

3.1.1 Permitted physicochemical modifications and antimicrobial treatments for the waters defined by origin

Waters defined by origin must not, prior to packaging, be modified or subjected to treatments other than those described in subsections below with the proviso that these modifications or treatments and the processes[2] used to achieve them do not change the essential physicochemical characteristics nor compromise the chemical, radiological and microbiological safety of these waters when packaged:

3.1.1.1 Selective treatments that modify the original composition:

- reduction and/or elimination of dissolved gases (and resulting possible change in pH);
- addition of carbon dioxide (and resulting change in pH) or re-incorporation of the original carbon dioxide present at emergence;
- reduction and/or elimination of unstable constituents such as iron, manganese, sulphur (as S^0 or S^{--}) compounds and carbonates in excess, under normal conditions of temperature and pressure, of the calco-carbonate equilibrium;
- addition of air, oxygen or ozone on condition that the concentration of by-products resulting from the ozone treatment is below the tolerance established under section 3.2.1;
- decrease and/or increase in temperature;
- reduction and/or separation of elements originally present in excess of maximum concentrations or of maximum levels of radioactivity set according to section 3.2.1.

3.1.1.2 Antimicrobial treatments for the waters defined by origin

Antimicrobial treatments may be used singly or in combination solely in order to conserve the original microbiological fitness for human consumption, original purity and safety of waters defined by origin.

3.1.2 Physical and chemical modifications and antimicrobial treatments for prepared waters

Prepared waters can be subjected to any microbial treatments and any treatments that modify the physical and chemical characteristics of the original water on condition that such treatments result in prepared waters that comply with all provisions of section 3.2 and 4 regarding the chemical, microbiological and radiological safety requirements for prepackaged waters.

---

[2] These processes include the techniques listed in Section 4.1 of the *Code of Hygienic Practice for Bottled/Packaged Drinking Waters (Other Than Natural Mineral Waters)* with the proviso that such techniques comply with the provisions outlined in Section 3.2.1 of the present standard.

3.2 Chemical and radiological quality of packaged waters

3.2.1 Health-related limits for chemical and radiological substances

No packaged water shall contain substances or emit radioactivity in quantities that may be injurious to health. To this effect, all packaged water shall comply with the health-related requirements of the most recent *"Guidelines for Drinking Water Quality"* published by the World Health Organization.

3.2.2 Addition of minerals

Any addition of minerals to water before packaging must comply with the provisions outlined in the present standard and, where applicable, with the provisions in *the Codex General Standard for Food Additives* (CODEX STAN 192-1995) and/or the *Codex General Principles for the Addition of Essential Nutrients to Foods* (CAC/GL 9-1987).

4. HYGIENE

4.1 Code of practice

It is recommended that all waters covered by the provisions of this standard be collected, transported, stored, and if applicable treated, and packaged in accordance with the *Recommended International Code of Practice – General Principles of Food Hygiene* (CAC/RCP 1-1991) and in accordance with *the Code of Hygienic Practice for Bottled/Packaged Drinking Waters (other than Natural Mineral Waters)* (CAC/RCP 48-2001).

4.2 Approval and inspection of the source for waters defined by origin

Initial approval or inspection of the source of waters defined by origin should be based upon appropriate scientific study adapted to the type of resource (hydrogeology, hydrology, etc.) and based on field survey of the source and of the recharge zone that shall demonstrate the safety of the source, the facilities and collection operations. The initial inspection of the source must be confirmed on a regular basis by periodic monitoring of the essential constituents, temperature, flow (in the case of natural springs) and the chemical and radiological factors specified under section 3.2.1 and the microbiological standards established in conformity with the latest *"Guidelines for Drinking Water Quality"* published by the World Health Organization. The results of source inspection should be made available to the importing country upon request.

5. LABELLING REQUIREMENTS

In addition to the *Codex General Standard for the Labelling of Prepackaged Foods* (CODEX STAN 1-1985), the following provisions shall apply:

5.1 Name of the product

Countries may select appropriate names for products, to be specified in national legislation, that reflect local consumer expectations arising from cultural and traditional practices.

However, in establishing such labelling requirements, consideration should be given to ensuring that any product complying with this standard may be represented in a way that reflects its classification within the standard and that consumers are not misled.

5.1.1 The name of the product shall be as follows, depending on its classification in accordance with Section 2.1.

5.1.1.1 **Waters defined by origin**

Any appropriate name (or names) in the case of waters that comply with the criteria described under section 2.1.1 and that meet additional criteria established by each country including restricting the name of such water to certain names or only one name. In the case of blends or mixtures of waters from different environment resources, each resource shall be labelled.

Only waters defined by origin, in accordance with the present standard, can be represented by names that refer to the origin or give an impression of specific origin. The names used or chosen by the countries, in accordance with the present standard, to represent prepared waters cannot apply to waters defined by origin and vice versa. When applicable, the additional criteria established by the countries for the definition of the chosen names cannot contravene the provisions of the present standard.

5.1.1.2 **Prepared waters**

Any appropriate name (or names) to designate prepared waters described under section 2.1.2 and that meet additional criteria established by each country including restricting the name of such water to certain names or only one name.

5.1.2 **Carbonation**

5.1.2.1 The following respective declarations should appear on the label in accordance with the following criteria:

In the case of ground waters defined by origin, "*naturally carbonated*" or "*naturally sparkling*" if, after packaging, carbon dioxide spontaneously and visibly is given off under normal conditions of temperature and pressure and the carbon dioxide originates from the source at emergence and is present at the same level as was present originally at emergence, with a possible re-incorporation of gas from the same source, taking into consideration a technical tolerance of ± 20%.

In the case of ground waters defined by origin, "*fortified with carbon dioxide*" if, after packaging, carbon dioxide spontaneously and visibly is given off under normal conditions of temperature and pressure and the carbon dioxide originates from the source at emergence but is present at a level at least 20% higher than the quantity present originally at emergence, with a possible re-incorporation of gas from the same source.

In the case of all waters, "*carbonated*" or "*sparkling*" if, after packaging, carbon dioxide spontaneously and visibly is given off under normal conditions of temperature and pressure and the carbon dioxide does not entirely originate from the same source as that of the water at emergence.

5.1.2.2 Words such as *"non carbonated"* or *"non sparkling"* or *"still"* may apply if, after packaging, there is no visible and spontaneous release of carbon dioxide under normal conditions of temperature and pressure when the packaged is opened.

5.2 Additional labelling requirements

5.2.1 Chemical composition

The total dissolved solid content of packaged waters may be declared on the principal display panel. With regard to waters defined by origin, the chemical composition that confers the characteristics to the product may also be declared on the label.

5.2.2 Geographic location

Where required by the authorities having jurisdiction, the precise geographic location of the specific environmental resource and/or the source of a water defined by origin must be declared in the manner prescribed in the applicable legislation.

5.2.3 Prepared water from a water distribution system

When prepared water is supplied by a public or private tap water distribution system and subsequently packaged/bottled, but has not undergone further treatment that would modify its original composition or to which carbon dioxide or fluoride have been added, the wording *"From a public or private distribution system"* must appear on the label along with the name of the product on the principal display panel.

5.2.4 Treatments

Where required by the authorities having jurisdiction, if a packaged/bottled water has been modified by a permitted treatment before packaging, the modification or the result of the treatment must be declared on the label in a manner prescribed in the applicable legislation.

5.3 Labelling prohibitions

5.3.1 No claims concerning medicinal (preventive, alleviative or curative) effects shall be made in respect of the properties of the product covered by this standard. Claims of other beneficial effects related to the health of the consumer shall not be made unless true and not misleading.

5.3.2 The name of the locality, hamlet or specified place may not form part of the trade name unless it refers to a water defined by origin collected at the place designated by that trade name.

5.3.3 The use of any statement or of any pictorial device which may create confusion in the mind of the public or in any way mislead the public about the nature, origin, composition and properties of packaged waters put on sale is prohibited.

6. METHODS OF ANALYSIS AND SAMPLING

See relevant Codex texts on methods of analysis and sampling.

CODE OF HYGIENIC PRACTICE FOR BOTTLED/ PACKAGED DRINKING WATERS
(Other than Natural Mineral Waters)

CAC/RCP 48-2001

CODE OF HYGIENIC PRACTICE FOR BOTTLED/ PACKAGED DRINKING WATERS (Other than Natural Mineral Waters)

CAC/RCP 48-2001

INTRODUCTION

International trade in bottled water has increased in recent years, both in quantity and diversity. Because of greater transport capacity, it is now possible to distribute bottled water not just as ship, rail, and road cargo but also as airfreight, the latter being used mainly in crisis situations due to the higher cost. By all these means of transport, a remedy for water shortages has become available when local water supply systems fail due to natural causes (such as droughts and earthquakes) or societal disasters (such as sieges or sabotage) and bottled water, both natural mineral water and diverse other sorts, has been brought in to meet such emergencies.

Aside from water shortages, real and perceived needs to improve health also have contributed to an escalating trade in bottled water. Increasingly it has been recognized that traditional suppliers of drinking water such as public and private waterworks may not always be able to guarantee the microbiological, chemical and physical safety of their product to the extent previously thought possible.

The contamination of water with viruses and parasitic protozoa is a serious concern to all consumers, particularly the immunocompromised. These pathogens are difficult to detect and bacterial indicators of their potential presence are not always reliable. Therefore it may be helpful to consumers to supply information regarding control measures the water has received. Protection of natural resources and such treatments as boiling, pasteurization, distillation, reverse osmosis filtration, absolute one micron or submicron filtration are some of the control measures used to guard against, inactivate or remove possible water contaminants such as oocysts of *Cryptosporidium parvum*, *Cyclospora cayentanensis*, and *Toxoplasma gondii* and cysts of other waterborne parasitic protozoa such as *Giardia* (*lamblia*) *intestinalis*, and *Entamoeba histolytica.*

It may be necessary that bottled drinking water products of particular chemical composition provide information concerning their proper consumption and/or have directions regarding whether or not they are suitable for infants and for the rehydration of infant formula.

SECTION 1. SCOPE, USE AND DEFINITIONS

1.1 Scope

This Code recommends general techniques for collecting, processing, packaging, storing, transporting, distributing, and offering for sale a variety of drinking waters (other than natural mineral water) for direct consumption. Recommendations concerning natural mineral water are provided in a separate Code (*Recommended International Code of Hygienic Practice for the Collecting, Processing and Marketing of Natural Mineral Waters* (CAC/RCP 33-1985)). All bottled/packaged drinking waters other than natural mineral water are covered by this Code.

1.2 Use of the document

It is emphasized that this document must be used in combination with the *Recommended International Code of Practice – General Principles of Food Hygiene*, (CAC/RCP 1-1969), including the HACCP Annex, whose paragraph numbers and section headings it maintains, supplementing or specifically applying them to bottled drinking waters (other than natural mineral waters). This Code should also be used in combination with the *Principles for the Establishment of and Application of Microbiological Criteria for Foods* (CAC/GL 21-1997).

1.3 Definitions

These definitions are supplemental to the definitions in section 2.3 of the *Recommended International Code of Practice – General Principles of Food Hygiene* (CAC/RCP 1-1969). The *Food*, and *Food handling* definitions apply only to this Code.

Bottled/packaged drinking water – Water filled into hermetically sealed containers of various compositions, forms, and capacities that is safe and suitable for direct consumption without necessary further treatment. Bottled drinking water is considered a food. The terms "drinking" and "potable" are used interchangeably in relation to water.

Drinking water systems – Public or private systems providing the consumer with tap water safe and suitable for direct consumption.

Establishment – Any suitable building(s), area(s) or surroundings in which water intended for bottling is collected, processed and bottled.

Food – For the purposes of this Code, the term includes bottled/packaged drinking water.

Food handling – Any operation pertaining to collecting, processing, bottling, packing of bottles, storing, transporting, distributing and marketing of bottled drinking water.

Ground water – Waters such as spring water, artesian water, and well water originating from subsurface aquifers. Ground waters may be classified broadly as protected or unprotected water. Protected ground waters are not directly influenced by surface water or the surface environment.

Ingredient – Any substance, including food additives, used to manufacture or prepare foods, intentionally added to a finished product, sometimes in a modified form (it may or may not be safe and suitable for human consumption without further treatment).

Surface water – Waters open to the atmosphere such as streams, rivers, lakes, ponds and reservoirs.

SECTION 2. PRIMARY PRODUCTION

These guidelines are supplemental to those set forth in Section III of the *Recommended International Code of Practice – General Principles of Food Hygiene*, (CAC/RCP 1-1969).

Prior to using a water resource for bottling purposes, its chemical composition and microbiological safety should be established over an appropriate period to allow for variations.

2.1 Environmental hygiene

2.1.1 Precautions in selecting a resource site

Hydrogeological data should determine the watershed and the perimeter (area surrounding the body of water from which supplies are drawn or the water's point of origin in the ground) that can be sources of contamination. These critical areas should be protected as much as possible.

All possible precautions should be taken within the protected perimeter (zone of protection) to avoid any pollution of, or external influence on, the quality of the ground or surface water. Disposal of liquid, solid or gaseous waste that could pollute the ground or surface water should be controlled. Disposal of pollutants such as microorganisms, fertilizers, hydrocarbons, detergents, pesticides, phenolic compounds, toxic metals, radioactive substances and other soluble organic and inorganic substances in the watershed should be avoided. Nor should drinking water resources be in the path of potential sources of underground contamination, such as sewers, septic tanks, industrial waste ponds, gas or chemical tanks, pipelines and solid waste disposal sites.

2.2 Hygienic production of water supplies

2.2.1 Protection of ground water supplies

2.2.1.1 Considerations for ground water supplies

It is not easy to distinguish between protected and unprotected ground water. Ground water supplies should be tested regularly for constancy of biological (including microbial), chemical, physical and, where necessary, radiological characteristics. The frequency of testing is determined by the hydrogeological evaluation, the amount of water collected, and the historical constancy pattern of a particular water supply. If contamination is detected, production of bottled water should cease until the water characteristics have returned to established parameters. Any underground supply from which water is collected, should be approved by an official authority having jurisdiction or by a third party with expertise for approving such underground supplies.

2.2.2 Protection of surface water supplies

Surface waters intended for bottling should be protected from contamination to the fullest extent possible even when treatments follow. Surface waters may be highly variable, so supplies should be tested frequently.

2.2.2.1 Considerations for surface water supplies

Stringency in determining which surface waters are suitable for bottling should be the rule, even when treatment(s) is foreseen.

2.3 Handling, storage and transport of water intended for bottling

2.3.1 Hygienic extraction or collection of water

2.3.1.1 At point of origin

The extraction or collection of water intended for bottling should be conducted in such a manner as to prevent other than the intended water from entering the extraction or collection device. The extraction or collection of water intended for bottling should also be conducted in a hygienic manner to prevent any contamination. Where sampling points are necessary, they should be designed and operated to prevent any contamination of the water.

2.3.1.2 Protection of the area of origin

The immediate surroundings of the extraction or collection area should be protected by limiting access to only authorized persons. Wellheads and spring outflows should be protected by a suitable structure to prevent entry by unauthorized individuals, pests, dust and other sources of contamination such as extraneous matter, drainage, floodwaters, and infiltration water.

2.3.1.3 Maintenance of extraction or collection facilities

Methods and procedures for maintaining the extraction facilities should be hygienic. They should not be a potential hazard to humans or a source of contamination for the water. Wells should be properly disinfected following construction and development of new wells nearby, after pump repair or replacement, or any well maintenance activity such as testing for and finding indicator organisms, pathogens, or abnormal plate counts in the water, and whenever biological growth inhibits proper operation. Water collection chambers should be disinfected within a reasonable time before use. Extraction devices such as those used for bore holes should be constructed and maintained in a manner that avoids contamination of the water and minimizes hazards to human health.

2.3.2 Storage and transport of water intended for bottling

When storage and transport of the water intended for bottling from the point of origin to the processing plant is necessary, these operations must be conducted in a hygienic manner to prevent any contamination.

In addition, see 2.3.2.1 and 2.3.2.2 below. Guidelines that are supplemental to those set forth in Section 3 of the *Recommended International Code of Practice – General Principles of Food Hygiene*, (CAC/RCP 1-1969) are found in the *Code of Hygienic Practice for Bulk Transport of Food and Semi-Packaged Foodstuffs* (CAC/RCP 47-2001). Directing the supply of water through piping from the point of origin wherever possible is one of the means of avoiding risks of contamination from bulk transport.

2.3.2.1 Requirements

Where or when they are necessary, bulk containers and conveyances such as tanks, pipings and tanker trucks should be designed and constructed so that they:

- do not contaminate the water intended for bottling;
- can be effectively cleaned and disinfected;

- provide effective protection from contamination, including dust and fumes; and
- allow any situation that arises to be checked easily.

2.3.2.2 **Use and maintenance**

Means of transport of water intended for bottling should be kept in an appropriate state of cleanliness, repair and condition. Containers and conveyances, particularly in bulk transport, should preferably be used only for transporting water intended for bottling. When this cannot be achieved, conveyances and bulk containers should be used exclusively for food transportation and must be cleaned and disinfected as necessary to prevent contamination. See also *Code of Hygienic Practice for the Transport of Bulk and Semi-Packed Food* (CAC/RCP 47-2001).

SECTION 3. ESTABLISHMENT: DESIGN AND FACILITIES

These guidelines are supplemental to those set forth in Section 4 of *the Recommended International Code of Practice – General Principles of Food Hygiene*, (CAC/RCP 1-1969).

3.1 Premises and rooms

In those areas of the processing establishment where containers are exposed to the external environment (*i.e.*, on the loading dock), especially prior to filling and sealing, specific preventive measures should be incorporated into the facility's design to avoid contamination of the containers used for bottled water.

3.2 Facilities

3.2.1 **Water supply not intended for bottling**

This section pertains to water for cleaning and disinfection purposes; not to water for bottling.

Water intended for bottling should be carried in completely separate lines from water not intended for bottling. These lines should be identified, preferably by different colours. There must be no cross-connections. Water used for cleaning and disinfection, should be potable (the standards of potability should not be less than those contained in the latest edition of the WHO *Guidelines for Drinking Water Quality*) if there is a chance that it comes into direct or indirect contact with water that is intended for bottling; otherwise it may be non-potable (if used where there is no direct or indirect contact with water for bottling). For storage, the provisions in the *Recommended International Code of Practice – General Principles of Food Hygiene* (CAC/RCP 1-1969) apply.

SECTION 4. ESTABLISHMENT: CONTROL OF OPERATION

These guidelines are supplemental to those set forth in Section V of *the Recommended International Code of Practice – General Principles of Food Hygiene*, (CAC/RCP 1-1969).

Water is an excellent vehicle for carrying substances in soluble, dispersed or emulsified form. Control measures must be taken at all steps of processing to ensure that food safety and suitability are not compromised by hazards or other contaminants during operations.

4.1 Key aspects of hygiene control systems

Waters, from drinking water systems, intended for bottling should meet all public drinking water standards (*i.e.*, chemical, microbiological, physical, radiological) established by the official authority having jurisdiction. For documentation of an approved source, firms using waters from drinking water systems may use drinking water system testing results showing full compliance with drinking water standards established by the official authority having jurisdiction in accordance with the *Guidelines for Drinking Water Quality* (WHO).

No waters intended for bottling should be accepted by an establishment if it is known to contain pathogens or excessive residues of pesticides or other toxic substances.

Water intended for bottling should be such (*i.e.* microbiologically, chemically, physically, and radiologically), that treatment if necessary (including multiple barrier treatments such as combination of filtration, chemical disinfection, etc.) of that water during processing results in finished bottled drinking water products that are safe and suitable for consumption. Generally, the higher the quality of the water intended for bottling, the less treatment is required to produce safe bottled drinking water products. Surface waters should be tested for safety frequently and treated as necessary.

A hazard analysis which takes into consideration pathogens and toxic substances should be undertaken in the overall context of the application of principles such as HACCP to the production of bottled water. This should provide the basis for determining the appropriate combination of control measures to reduce, eliminate or prevent, as necessary, hazards (microbiological, chemical and radiological) for the production of safe bottled water. Waters originating from protected underground supplies are less likely to require treatment than waters originating from surface supplies or unprotected underground supplies.

When necessary, treatment of waters intended for bottling, to reduce, remove or prevent growth of pathogens, may include the application of chemical processes (such as chlorination, ozonation, carbonation) and physical agents or processes (such as high heat, ultraviolet radiation, filtration). These treatments can be used singly or in combination as multiple barriers. Treatments vary in their effectiveness against specific organisms. Bottled waters produced with the use of an adequate multiple barrier treatment technique will be less likely to contain pathogens of public health concern.

When necessary, treatments to remove or reduce chemical substances may include chemical and particulate (mechanical) filtration such as achieved with surface filters (e.g., pleated membrane filters) or depth filters (e.g., sand or compressed fibre (cartridge) filters), activated carbon filtration, demineralization (deionization, water softening, reverse osmosis, nano-filtration), and aeration. These treatments for chemicals may not adequately reduce or remove microorganisms and, likewise, treatments for microorganisms may not adequately reduce or remove chemicals and particulate matters.

All treatments of water intended for bottling should be carried out under controlled conditions to avoid any type of contamination, including the formation of toxic by-products (particularly bromates) and the presence of residues of water treatment chemicals in amounts that raise health concerns in accordance with relevant WHO guidelines.

4.2 Packaging

The requirements in the *Recommended International Code of Practice – General Principles of Food Hygiene* (CAC/RCP 1-1969) cover these topics.

4.2.1 Washing and disinfecting of containers

Reused containers and where necessary other containers should be washed and disinfected in an appropriate system and positioned within the processing plant so as to minimize post-sanitizing contamination prior to filling and sealing. Disposable containers may be ready for use without prior washing and disinfecting. Determine if this is the case; if not, treat as carefully as reusable containers.

4.2.2 Filling and sealing of containers

Bottling operations (*i.e.* filling and sealing of containers) should be conducted in a manner that protects against contamination. Control measures include the use of an enclosed area and a containment enclosed system separate from other operations of the processing plant to protect against contamination. Dust, dirt, microorganisms in the air, and condensation should be controlled and monitored.

4.2.3 Product containers and closures

Reusable containers should not have been used for any purpose that may lead to contamination of the product and should be individually inspected for suitability. New containers should be inspected and, if necessary, cleaned and disinfected.

4.2.4 Use of closures

Closures are generally supplied in a ready to use state and should be tamper resistant; they are not reusable.

SECTION 5. ESTABLISHMENT: MAINTENANCE AND SANITATION

The requirements in the *Recommended International Code of Practice – General Principles of Food Hygiene* (CAC/RCP 1-1969) cover these topics.

SECTION 6. ESTABLISHMENT: PERSONAL HYGIENE

The requirements in the *Recommended International Code of Practice – General Principles of Food Hygiene* (CAC/RCP 1-1969) cover this topic.

SECTION 7. TRANSPORTATION AND STORAGE OF BOTTLED WATER

Guidelines that are supplemental to those set forth in Section 8 of the *Recommended International Code of Practice – General Principles of Food Hygiene* (CAC/RCP 1-1969) are found in the *Code of Hygienic Practice for the Transport of Foodstuffs in Bulk and Semi-Packed Food* (CAC/RCP 47-2001). For storage, the provisions in the *Recommended International Code of Practice – General Principles of Food Hygiene* (CAC/RCP 1-1969) apply.

SECTION 8. PRODUCT INFORMATION AND CONSUMER AWARENESS

These requirements are covered in the *Recommended International Code of Practice – General Principles of Food Hygiene*, (CAC/RCP 1-1969). See also the text in the Introduction of this document.

SECTION 9. TRAINING

The requirements made in the *International Recommended Code of Practice – General Principles of Food Hygiene* (CAC/RCP 1-1969) cover this topic.

APPENDIX 1

MICROBIOLOGICAL AND OTHER SPECIFICATIONS

Section 5.2.3 Microbiological and Other Specifications of the *International Recommended Code of Practice – General Principles of Food Hygiene*, (CAC/RCP 1-1969) applies.

The provisions of the *Principles for the Establishment and Application of Microbiological Criteria for Foods* (CAC/GL 21-1997) apply.

Microbiological and additional specifications for bottled drinking waters (other than natural mineral waters) are those of the World Health Organization *Guidelines for Drinking Water Quality*.

RECOMMENDED INTERNATIONAL CODE OF HYGIENIC PRACTICE FOR THE COLLECTING, PROCESSING AND MARKETING OF NATURAL MINERAL WATERS

CAC/RCP 33-1985

RECOMMENDED INTERNATIONAL CODE OF HYGIENIC PRACTICE FOR THE COLLECTING, PROCESSING AND MARKETING OF NATURAL MINERAL WATERS

CAC/RCP 33-1985

SECTION 1. SCOPE

This Code recommends appropriate general techniques for collecting natural mineral water, its treatment, bottling, packaging, storage, transport, distribution and sale for direct consumption, so as to guarantee a safe, healthy and wholesome product.

SECTION 2. DEFINITIONS

2.1 For the purpose of this code the following expressions have the meaning stated:

2.1.1 ***Natural mineral waters*** – all waters meeting the requirement of the Standard for Natural Mineral Waters (CODEX STAN 108-1981).

2.1.2 ***Adequate*** – sufficient to accomplish the intended purpose of this code.

2.1.3 ***Cleaning*** – the removal of soil, food residues, dirt, grease or other objectionable matter.

2.1.4 ***Contamination*** – the occurrence of any objectionable matter in the product.

2.1.5 ***Disinfection*** – the reduction, without adversely affecting the natural mineral water, by means of hygienically satisfactory chemical agents and/or physical methods, of the number of microorganisms to a level that will not lead to harmful contamination of natural mineral water.

2.1.6 ***Establishment*** – any building(s) or areas in which natural mineral water is handled after collection and the surroundings under the control of the same management.

2.1.7 ***Handling of natural mineral water*** – any manipulation with regard to collecting, treating, bottling, packaging, storing, transport, distribution and sale of natural mineral water.

2.1.8 ***Food Hygiene*** – all measures necessary to ensure the safety, soundness and wholesomeness of natural mineral water at all stages from its exploitation and processing until its final consumption.

2.1.9 ***Packaging Material*** – any containers such as cans, bottles, cartons, boxes, cases or wrapping and covering material such as foil, film, metal paper and wax-paper.

2.1.10 ***Pests*** – any animals capable of directly or indirectly contaminating natural mineral water.

2.1.11 ***Containers*** – any bottle, carton, can or other container to be filled with natural mineral water, properly labelled and intended for sale.

2.1.12 ***Aquifers*** – any solid permeable mass of rocks (layer) containing natural mineral water.

2.1.13 ***Spring*** – any natural mineral water discharging genuinely from the ground.

SECTION 3. PRESCRIPTIONS OF THE RESOURCES OF NATURAL MINERAL WATERS

A. PROTECTION OF ALIMENTARY RESERVOIRS AND AQUIFERS

3.1 Authorization

Any spring, well or drilling intended for the collection of natural mineral water should be approved by the official authority having jurisdiction for this region.

3.2 Determination of the genesis of natural mineral water

As far as it is methodologically possible in each case, a precise analysis should be carried out on the origin of natural mineral waters, the period of their residence in the ground before being collected and their chemical and physical qualities.

3.3 Perimeter of protection

If possible areas wherein natural mineral water might be polluted or its chemical and physical qualities otherwise deteriorated should be determined by a hydrologist. Where indicated by hydrogeological conditions and considering the risks of pollution and physical, chemical and biochemical reactions, several perimeters with separate dimensions may be provided for.

3.4 Protective measures

All possible precautions should be taken within the protected perimeters to avoid any pollution of, or external influence on, the chemical and physical qualities of natural mineral water.

It is recommended that regulations be established for the disposal of liquid, solid or gaseous waste, the use of substances that might deteriorate natural mineral water (e.g. by agriculture) as well as for any possibility of accidental deterioration of natural mineral water by natural occurrences such as a change in the hydrogeological conditions. Particular consideration should be given to the following potential pollutants: bacteria, viruses, fertilizers, hydrocarbons, detergents, pesticides, phenolic compounds, toxic metals, radioactive substances and other soluble organic or inorganic substances.

Even where nature provides apparently sufficient protection against surface pollution, potential hazards should be taken into consideration, such as mining, hydraulic and engineering facilities etc.

B. HYGIENE PRESCRIPTIONS FOR THE COLLECTION OF NATURAL MINERAL WATER

3.5 Extraction

The withdrawal of natural mineral water (from springs, galleries, genuine or drilled wells) must be performed in conformity with the hydrogeological conditions in such a manner as to prevent any other than the natural mineral water from entering or, should there be pumping facilities, prevent any extraneous water from entering by reducing the supply. The natural mineral water thus collected or pumped should be protected in such a way that it will be safe from pollution whether caused by natural occurrence or actions or neglect or ill will.

3.6 Materials

The pipes, pumps or other possible devices coming into contact with natural mineral water and used for its collection should be made of such material as to guarantee that the original qualities of natural mineral water will not be changed.

3.7 Protection of the extraction area

In the immediate surroundings of springs and wells, precautionary measures should be taken to guarantee that no pollutant whatsoever can enter the extraction area. The extraction area should be inaccessible to non-authorized people by providing adequate devices (e.g. enclosure). Any use not aiming at the collection of natural mineral water should be forbidden in this area.

3.8 The exploitation of natural mineral water

The condition of the extraction facilities, areas of extraction and perimeters of protection as well as the quality of the natural mineral water should periodically be checked. To control the stability of the chemical and physical particulars of the natural mineral water derived – besides the natural variations –automatic measurements of the typical characteristics of water should be carried out and notified (e.g., electrical conductance, temperature, content of carbon dioxide) or frequent partial analyses should be done.

C. MAINTENANCE OF EXTRACTION FACILITIES

3.9 Technical aspects

Methods and procedures for maintaining the extraction facilities should be hygienic and not be a potential health hazard to humans or a source of contamination to natural mineral water. From the hygiene standpoint, servicing of the extraction installations should meet the same standards as those required for the bottling or treatment.

3.10 Equipment and reservoirs

Equipment and reservoirs used for extraction of natural mineral water should be constructed and maintained in order to minimize all hazards to human health and to avoid contamination.

3.11 Storage at the point of extraction

The quantity of natural mineral water stored at the point of extraction should be as low as possible. The storing should furthermore guarantee protection against contamination or deterioration.

D. TRANSPORT OF NATURAL MINERAL WATER

3.12 Means of transport, piping and reservoirs

Any vehicle, piping or reservoir used in the processing of natural mineral water from its source to the bottling facilities, the latter included, should comply with the necessary requirements and be made of inert material such as ceramic and stainless steel which prevents any deterioration, be it by water, handling, servicing or disinfection; it should allow easy cleaning.

3.13 Maintenance of vehicles and reservoirs

Any vehicle or reservoir should be properly cleaned and if necessary disinfected and kept in good repair so as to not to present any danger of contamination to natural mineral water and of deterioration of the essential qualities of natural mineral water.

SECTION 4. ESTABLISHMENT FOR PROCESSING NATURAL MINERAL WATERS – DESIGN AND FACILITIES

4.1 Location

Establishments should be located in areas which are free from objectionable odours, smoke, dust or other contaminants and are not subject to flooding.

4.2 Roadways and areas used by wheeled traffic

Such roadways and areas serving the establishment which are within its boundaries or in its immediate vicinity should have a hard paved surface suitable for wheeled traffic. There should be adequate drainage and provision should be made for protection of the extraction area in accordance with sub-section 3.7 where appropriate and to allow for cleaning. Adequate road signals may be provided to call the attention of road users to the existence of a natural mineral water extraction area.

4.3 Buildings and facilities

4.3.1 Type of construction

Buildings and facilities should be of sound construction in accordance with the provisions of sub-section 3.7 and maintained in good repair.

4.3.2 Disposition of holding facilities

Rooms for recreation, for storing or packaging of raw material and areas for the cleaning of containers to be re-used should be apart from the bottling areas to prevent the end-product from being contaminated. Raw and packaging materials and any other additions which come into contact with natural mineral water should be stored apart from other material.

4.3.3 Adequate working space should be provided to allow for satisfactory performance of all operations.

4.3.4 The design should be such as to permit easy and adequate cleaning and to facilitate proper supervision of natural mineral water hygiene.

4.3.5 The buildings and facilities should be designed to provide separation by partition, location or other effective means between those operations which may cause cross-contamination.

4.3.6 Buildings and facilities should be designed to facilitate hygienic operations by means of a regulated flow in the process from the arrival of the natural mineral water at the premises to the finished product, and should provide for appropriate temperature conditions for the process and the product.

4.3.7 **Natural mineral water handling, storing and bottling areas**

Floors, where appropriate, should be of water-proof, non-absorbent, washable, non-slip and non-toxic materials, without crevices, and should be easy to clean and disinfect. Where appropriate, floors should slope sufficiently for liquids to drain to trapped outlets.

Walls, where appropriate, should be of water-proof, non-absorbent, washable and non-toxic materials and should be light coloured. Up to a height appropriate for the operation they should be smooth and without crevices, and should be easy to clean and disinfect. Where appropriate, angles between walls, between walls and floors, and between walls and ceilings should be sealed and coved to facilitate cleaning.

Ceilings should be so designed, constructed and finished as to prevent the accumulation of dirt and minimize condensation, mould development and flaking, and should be easy to clean.

Windows and other openings should be so constructed as to avoid accumulation of dirt and those which open should be fitted with screens. Screens should be easily movable for cleaning and kept in good repair. Internal window sills, if present, should be sloped to prevent use as shelves.

Doors should have smooth, non-absorbent surfaces and, where appropriate, be self-closing and close fitting.

Stairs, lift cages and auxiliary structures such as platforms, ladders, chutes, should be so situated and constructed as not to cause contamination to food. Chutes should be constructed with inspection and cleaning hatches.

Piping for natural mineral water lines should be independent of potable and non-potable waters.

4.3.8 In natural mineral water handling areas all overhead structures and fittings should be installed in such a manner as to avoid contamination directly or indirectly of natural mineral water and raw materials by condensation and drip, and should not hamper cleaning operations. They should be insulated where appropriate and be so designed and finished as to prevent the accumulation of dirt and to minimize condensation, mould development and flaking. They should be easy to clean.

4.3.9 Living quarters, toilets and areas where animals are kept should be completely separated from and should not open directly on to natural mineral water handling areas.

4.3.10 Where appropriate, establishments should be so designed that access can be controlled.

4.3.11 The use of material which cannot be adequately cleaned and disinfected, such as wood, should be avoided unless its use would clearly not be a source of contamination.

4.3.12 **Canalization, drainage lines**

Canalization and drainage and used water lines as well as any possible waste storage area within the protected perimeter should be built and maintained in such a manner as not to present any risk whatsoever of polluting aquifers and springs.

4.3.13 **Fuel storage area**

Any storage area or tank for the storing of fuels such as coal or hydrocarbons should be designed, protected, controlled and maintained in such a manner as not to present a risk of aquifers and springs being polluted during the storage and manipulation of these fuels.

4.4 Hygienic facilities

4.4.1 Water supply

4.4.1.1 An ample supply of *potable* water in compliance with Section 5.5.1 of the *Codex Code of Practice – General Principles of Food Hygiene* (CAC/RCP 1-1969) under adequate pressure and of suitable temperature should be available with adequate facilities for its storage, where necessary, and disstribution, and with adequate protection against contamination. The standards of potability should not be less than those contained in the latest edition of "*Guidelines for Drinking-water Quality*" (WHO).

4.4.1.2 Natural mineral water, potable water, non potable water for steam production or for refrigeration or any other use should be carried in completely separate lines with no cross connection between them and without back siphonage. It would be desirable that these lines be identified by different colours. Steam used in direct contact with natural mineral water and natural mineral water contact surfaces should contain no substances which may be hazardous to health or may contaminate the food.

4.4.2 **Effluent and waste disposal**

Establishments should have an efficient effluent and waste disposal system which should at all times be maintained in good order and repair. All effluent lines (including sewer systems) should be large enough to carry peak loads and should be so constructed as to avoid contamination of potable water supplies.

4.4.3 **Changing facilities and toilets**

Adequate, suitable and conveniently located changing facilities and toilets should be provided in all establishments. Toilets should be so designed as to ensure hygienic removal of waste matter. These areas should be well lit, ventilated and where appropriate heated, and should not open directly on to natural mineral water handling areas. Hand washing facilities with warm or hot and cold water, a suitable hand-cleaning preparation, and with suitable hygienic means of drying hands, should be provided adjacent to toilets and in such a position that the employee must pass them when returning to the processing area. Where hot and cold water are available mixing taps should be provided. Where paper towels are used, a sufficient number of dispensers and receptacles should be provided near to each washing facility. Care should be taken that these receptacles for used paper towels are regularly emptied. Taps of a non-hand operable type are desirable. Notices should be posted directing personnel to wash their hands after using the toilet.

4.4.4 **Hand washing facilities in natural mineral water processing areas**

Adequate and conveniently located facilities for hand washing and drying should be provided wherever the process demands. Where appropriate, facilities for hand disinfection should also be provided. Warm or hot and cold water and a suitable hand-cleaning preparation should be provided. Where hot and cold water are available mixing taps should be provided. There should be suitable hygienic means of drying hands. Where paper towels are used, a sufficient number of dispensers and receptacles should be provided adjacent to each washing facility. Taps of a non-hand operable type are desirable. The facilities should be furnished with properly trapped waste pipes leading to drains.

4.4.5 **Disinfection facilities**

Where appropriate, adequate facilities for cleaning and disinfection of working implements and equipment should be provided. These facilities should be constructed of corrosion resistant materials, capable of being easily cleaned, and should be fitted with suitable means of supplying hot and cold water in sufficient quantities.

4.4.6 **Lighting**

Adequate natural or artificial lighting should be provided throughout the establishment. Where appropriate, the lighting should not alter colours and the intensity should not be less than:

- 540 lux (50 foot candles) at all inspection points
- 220 lux (20 foot candles) in work rooms
- 110 lux (10 foot candles) in other areas.

Light bulbs and fixtures suspended over natural mineral water in any stage of production should be of a safety type and protected to prevent contamination of natural mineral water in case of breakage.

4.4.7 **Ventilation**

Adequate ventilation should be provided to prevent excessive heat, steam condensation and dust and to remove contaminated air. The direction of the air flow should never be from a dirty area to a clean area. Ventilation openings should be provided with a screen or other protecting enclosure of non-corrodible material. Screens should be easily removable for cleaning.

4.4.8 **Facilities for storage of waste and inedible material**

Facilities should be provided for the storage of waste and inedible material prior to removal from the establishment. These facilities should be designed to prevent access to waste or inedible material by pests and to avoid contamination of natural mineral water, potable water, equipment, buildings or roadways on the premises.

4.5 Equipment and utensils

4.5.1 **Materials**

All equipment and utensils used in natural mineral water handling areas and which may contact the natural mineral water should be made of material which does not transmit toxic substances, odour or taste, is non-absorbent, is resistant to corrosion and is capable of withstanding repeated cleaning and disinfection. Surfaces should be smooth and free from pits and crevices. The use of wood and other materials which cannot be adequately cleaned and disinfected should be avoided except when their use would clearly not be a source of contamination. The use of different materials in such a way that contact corrosion can occur should be avoided.

4.5.2 **Hygienic design, construction and installation**

4.5.2.1 All equipment and utensils should be so designed and constructed as to prevent hygienic hazards and permit easy and thorough cleaning and disinfection.

SECTION 5. ESTABLISHMENT: HYGIENE REQUIREMENTS

5.1 Maintenance

The buildings, equipment, utensils and all other physical facilities of the establishment, including drains, should be maintained in good repair and in an orderly condition. As far as practicable, rooms should be kept from steam, vapour and surplus water.

5.2 Cleaning and disinfection

5.2.1 Cleaning and disinfection should meet the requirements of this Code. For further information on cleaning and disinfection procedures see Annex I, Revised *Recommended International Code of Practice – General Principles of Food Hygiene* (CAC/RCP 1-1969).

5.2.2 To prevent contamination of natural mineral water, all equipment and utensils should be cleaned as frequently as necessary and disinfected whenever circumstances demand.

5.2.3 Adequate precautions should be taken to prevent natural mineral water from being contaminated during cleaning or disinfection of rooms, equipment or utensils, by water and detergents or by disinfectants and their solutions. Detergents and disinfectants should be suitable for the purpose intended and should be acceptable to the official agency having jurisdiction. Any residues of these agents on a surface which may come in contact with natural mineral water should be removed by thorough rinsing with water in compliance with Section 5.5.1 of the *Recommended International Code of Hygienic Practice – General Principles of Food Hygiene* (CAC/RCP 1-1969) before the area or equipment is again used for handling natural mineral water.

5.2.4 Either immediately after cessation of work for the day or at such other times as may be appropriate, floors, including drains, auxiliary structures and walls of natural mineral water handling areas should be thoroughly cleaned.

5.2.5 Changing facilities and toilets should be kept clean at all times.

5.2.6 Roadways and yards in the immediate vicinity of and serving the premises should be kept clean.

5.3 Hygiene control programme

A permanent cleaning and disinfection schedule should be drawn up for each establishment to ensure that all areas are appropriately cleaned and that critical areas, equipment and material are designated for special attention. A single individual, who should preferably be a permanent member of the staff of the establishment and whose duties should be independent of production, should be appointed to be responsible for the cleanliness of the establishment. He should have a thorough understanding of the significance of contamination and the hazards involved. All cleaning personnel should be well-trained in cleaning techniques.

5.4 Storage and disposal of waste

Waste material should be handled in such a manner as to avoid contamination of natural mineral water or potable water. Care should be taken to prevent access to waste by pests. Waste should be removed from the natural mineral water handling and other working areas as often as necessary and at least daily. Immediately after disposal of the waste, receptacles used for storage and any equipment which has come into contact with the waste should be cleaned and disinfected. The waste storage area should also be cleaned and disinfected.

5.5 Exclusion of animals

Animals that are uncontrolled or that could be a hazard to health should be excluded from establishments.

5.6 Pest control

5.6.1 There should be an effective and continuous programme for the control of pests. Establishments and surrounding areas should be regularly examined for evidence of infestation.

5.6.2 Should pests gain entrance to the establishment, eradication measures should be instituted. Control measures involving treatment with chemical, physical or biological agents should only be undertaken by or under direct supervision of personnel who have a thorough understanding of the potential hazards to health resulting from the use of these agents, including those hazards which may arise from residues retained in the natural mineral water, such measures should only be carried out in accordance with the recommendations of the official agency having jurisdiction.

5.6.3 Pesticides should only be used if other precautionary measures cannot be used effectively. Before pesticides are applied, care should be taken to safeguard natural mineral water, equipment and utensils from contamination. After application, contaminated equipment and utensils should be thoroughly cleaned to remove residues prior to being used again.

5.7 Storage of hazardous substances

5.7.1 Pesticides or other substances which may represent a hazard to health should be suitably labelled with a warning about their toxicity and use. They should be stored in locked rooms or cabinets used only for that purpose and dispensed and handled only by authorized and properly trained personnel or by persons under strict supervision of trained personnel. Extreme care should be taken to avoid contaminating natural mineral water.

5.7.2 Except when necessary for hygienic or processing purposes, no substance which could contaminate natural mineral water should be used or stored in natural mineral water handling areas.

5.8 Personal effects and clothing

Personal effects and clothing should not be deposited in natural mineral water handling areas.

SECTION 6. PERSONNEL HYGIENE AND HEALTH REQUIREMENTS

6.1 Hygiene training

Managers of establishments should arrange for adequate and continuing training of all natural mineral water handlers in hygienic handling of natural water and in personal hygiene so that they understand the precautions necessary to prevent contamination of natural mineral water. Instructions should include relevant parts of this Code.

6.2 Medical examination

Persons who come into contact with natural mineral water in the course of their work should have a medical examination prior to employment if the official agency having jurisdiction, acting on medical advice, considers that this is necessary, whether because of epidemiological considerations or the medical history of the prospective natural mineral water handler. Medical examination of natural mineral water handlers should be carried out at other times when clinically or epidemiologically indicated.

6.3 Communicable diseases

The management should take care to ensure that no person, while known or suspected to be suffering from, or to be a carrier of a disease likely to be transmitted through food or while afflicted with infected wounds, skin infections, sores or with diarrhoea, is permitted to work in any natural mineral water handling area in any capacity in which there is any likelihood of such a person directly or indirectly contaminating natural mineral water with pathogenic micro-organisms. Any person so affected should immediately report to the management that he is ill.

6.4 Injuries

Any person who has a cut or wound should not continue to handle natural mineral water or natural mineral water contact surfaces until the injury is completely protected by a waterproof covering which is firmly secured, and which is conspicuous in colour. Adequate first-aid facilities should be provided for this purpose.

6.5 Washing of hands

Every person, while on duty in a natural mineral water handling area, should wash his hands frequently and thoroughly with a suitable hand cleaning preparation under running warm water in compliance with Section 5.5.1 of the *Recommended International Code of Hygienic Practice – General Principles of Food Hygiene* (CAC/RCP 1-1969). Hands should always be washed before commencing work, immediately after using the toilet, after handling contaminated material and whenever else necessary. After handling any material which might be capable of transmitting disease, hands should be washed and disinfected immediately. Notices requiring hand-washing should be displayed. There should be adequate supervision to ensure compliance with this requirement.

6.6 Personal cleanliness

Every person engaged in a natural mineral water handling area should maintain a high degree of personal cleanliness while on duty, and should at all times while so engaged wear suitable protective clothing including head covering and footwear, all of which should be cleanable unless designed to be disposed of and should be maintained in a clean condition consistent with the nature of the work in which the person is engaged. Aprons and similar items should not be washed on the floor. During periods where natural mineral water is manipulated by hand, any jewellery that cannot be adequately disinfected should be removed from the hands. Personnel should not wear any insecure jewellery when engaged in natural mineral water handling.

6.7 Personal behaviour

Any behaviour which could result in contamination of natural mineral water, such as eating, use of tobacco, chewing (e.g. gum, sticks, betel nuts, etc.) or unhygienic practices such as spitting, should be prohibited in natural mineral water handling areas.

6.8 Visitors

Precautions should be taken to prevent visitors to natural mineral water handling areas from contaminating the product. These may include the use of protective clothing. Visitors should observe the provisions recommended in paragraphs 5.8, 6.3, 6.4 and 6.7 of this Code.

6.9 Supervision

Responsibility for ensuring compliance by all personnel with all requirements of Sections 6.1–6.8 inclusive should be specifically allocated to competent supervisory personnel.

SECTION 7. ESTABLISHMENT: HYGIENIC PROCESSING REQUIREMENTS

7.1 Raw material requirements

To guarantee a good and stable quality of natural mineral water, certain criteria should be monitored regularly, e.g.

7.1.1 Spring discharge, temperature of the natural mineral water;

7.1.2 Appearance of the natural mineral water;

7.1.3 Odour and taste of the natural mineral water;

7.1.4 The conductance of natural mineral water or any other adequate parameter;

7.1.5 The microbiological flora.

7.2 Should there be a perceptible lack in meeting the standards, the necessary corrective measures are immediately to be taken.

7.3 Treatment

The treatment may include decantation, filtration, airing and where necessary application of offtake of carbon dioxide ($C0_2$).

7.3.1 Processing should be supervised by technically competent personnel.

7.3.2 All steps in the production process, including packaging, should be performed without unnecessary delay and under conditions which will prevent the possibility of contamination, deterioration, or the development of pathogenic and spoilage micro-organisms.

7.3.3 Rough treatment of containers should be avoided to prevent the possibility of contamination of the processed product.

7.3.4 Treatment and necessary controls should be such as to protect against contamination or development of a public health hazard and against deterioration within the limits of good commercial practice.

7.3.5 All contaminated equipment which has been in contact with raw materials should be thoroughly cleaned and disinfected prior to being used in contact with the end-products.

7.4 Packaging material and containers

7.4.1 All packaging material should be stored in a clean and sanitary manner. The material should be appropriate for the product to be packed and for the expected conditions of storage and should not transmit to the product objectionable substances beyond the limits acceptable to the official agency having jurisdiction. The packaging material should be sound and should provide appropriate protection from contamination. Only packaging material required for immediate use should be kept in the packing or filling area.

7.4.2 Product containers should not have been used for any purpose that may lead to contamination of the product. Used containers, also new containers if there is a possibility that they have been contaminated, should be cleaned and disinfected. When chemicals are used for these purposes, the container should be rinsed as prescribed under 5.2.3. Containers should be well drained after rinsing. Used and, when necessary, unused containers should be inspected immediately before filling.

7.5 Filling and sealing of containers

7.5.1 Packaging should be done under conditions that preclude the introduction of contaminants into the product.

7.5.2 The methods, equipment and material used for sealing should guarantee a tight and impervious sealing and not damage the containers nor deteriorate the chemical, bacteriological and organoleptic qualities of natural mineral water.

7.6 Packaging of containers

The packaging of containers should protect the latter from contamination and damage and allow appropriate handling and storing.

7.7 Lot identification

Each container shall be permanently marked in code or in clear to identify the producing factory and the lot. A lot is a quantity of natural mineral water produced under identical conditions, all packages of which should bear a lot number that identifies the production during a particular time interval, and usually from a particular "line" or other critical processing unit.

7.8 Processing and production records

Permanent, legible and dated records of pertinent processing and production details should be kept concerning each lot. These records should be retained for a period that exceeds the shelf life of the product. Records should also be kept of the initial distribution by lot.

7.9 Storage and transport of the end-product

The end-product should be stored and transported under such conditions as will preclude contamination with and/or proliferation of micro-organisms and protect against deterioration of the product or damage to the container. During storage, periodic inspection of the end-product should take place to ensure that only natural mineral water which is fit for human consumption is despatched and that end-product specifications should be complied with when they exist.

7.10 Sampling and laboratory control procedure

The following are intended as guidelines for testing the water at the source and at critical control points:

Natural mineral water should contain no parasites and should meet the following criteria:

		n	c	m	M	Method
1.	Coliforms	5 x 250 ml	1	0	1*	ISO/DIS 9308/1
2.	Faecal streptococci	5 x 250 ml	1	0	1*	(ISO 7899/2)
3.	Spore-forming sulfite-reducing anaerobes	5 x 250 ml	1	0	1*	(ISO 6461/2)
4.	*Pseudomonas aeruginosa*	5 x 250 ml	0	0	–	(ISO 8360/2)
5.	Aerobic microbial count: The maximum permissible total aerobic counts per millilitre at 20–22°C and 37°C depend on the unique characteristics of the source and should be fixed by the authority having jurisdiction.					

* In cases of positive results (M ≥ 1) additional samples should be examined to determine the reason for positive results.

SECTION 8. END-PRODUCT SPECIFICATIONS

During marketing, natural mineral water:

(i) shall be of such a quality that it will not represent a risk to the health of the consumer (absence of pathogenic micro-organisms);

(ii) furthermore it shall be in conformity with the following microbiological specifications:

First examination		Decision		
Coliforms*	1 × 250 ml	if absent	→	accepted
Group D Streptococci	1 × 250 ml	if ≥ 1 or ≤ 2	→	second examination is carried out[1]
		if > 2	→	rejected
P. aeruginosa	1 × 250 ml	if absent	→	accepted
		if ≥ 1	→	rejected

Second examination (4 x 250 ml)	c[2]	m	M	Method
Coliforms*	1	0	2	ISO Methods[3]
Group D Streptococci	1	0	2	ISO Methods[3]
P. aeruginosa	0	0	0	ISO Methods[3]

[1] The second examination shall include detection of coliforms, Group D Streptococci, and *P. aeruginosa*.
[2] Results from the first and second examination.
[3] Methods to be elaborated.
* Shall not be *E. coli*.